YOUR KNOWLEDGE H

- We will publish your bachelor's and
  master's thesis, essays and papers

- Your own eBook and book -
  sold worldwide in all relevant shops

- Earn money with each sale

Upload your text at www.GRIN.com
and publish for free

Bibliographic information published by the German National Library:

The German National Library lists this publication in the National Bibliography; detailed bibliographic data are available on the Internet at http://dnb.dnb.de .

Imprint:

Copyright © 2016 GRIN Verlag, Open Publishing GmbH
Print and binding: Books on Demand GmbH, Norderstedt Germany
ISBN: 9783668304048

This book at GRIN:

http://www.grin.com/en/e-book/340642/artificial-blood-vessels-for-coronary-artery-disease-patients

Marvin Namanda

Artificial blood vessels for coronary artery disease patients

GRIN Publishing

GRIN - Your knowledge has value

Since its foundation in 1998, GRIN has specialized in publishing academic texts by students, college teachers and other academics as e-book and printed book. The website www.grin.com is an ideal platform for presenting term papers, final papers, scientific essays, dissertations and specialist books.

Visit us on the internet:

http://www.grin.com/

http://www.facebook.com/grincom

http://www.twitter.com/grin_com

ARTIFICIAL BLOOD VESSELS FOR CORONARY HEART DISEASE

This text was written by a non-native English speaker. Please excuse any errors or inconsistencies.

Content

CHAPTER ONE

1.1 Overview

Coronary artery disease is among the fetal disease affecting most individuals across the globe. It occurs when the coronary arteries are clogged by either clot or any substance that will lead to constriction of the lumen hence leading to stenosis. As such, patients with severe stenosis are required to undergo coronary artery bypass grafting surgery to enable efficient flow of blood to region supplied by the affected coronary artery. Coronary artery bypass surgery is an open chest operation that involves connecting the aorta to coronary artery after removal of the stenos using an artery or vein from another part of the body (Cohn, 2012).

It is preferred that a prosthetic arteries be used in the operation so as to avoid the complications that may result from the use of natural blood vessels. However, the use of a prosthetic artery has been associated with several cases of thrombus formation, poor cell growth, poor proliferations and poor adhesion. Moreover, the prosthetic artery fabrication to mimic the natural blood vessels has been reported to a challenge hence increasing chances of rejection (Fung, 2013). The purpose of this paper is to give a detailed analysis of artificial blood vessels manufacturing using techniques such as additive manufacturing (3D printing). Scaffolding designs will be studied using computer aided software and choice of bio-compatible material that matches requirements for manufacturing the artificial blood vessels will be discussed too. Finally, the paper will look into the mechanical properties of the scaffoldings designs (Fink & Helen, 2009).

CHAPTER TWO

2.1 Artificial Blood Vessel

Synthetic blood vessels were chemically produced during World War I by a French-American surgeon Alexis carrel to restore the blood circulation in the wounded soldiers. At this time, 1873, Alex was perfect in sewing ruptured blood vessels and this won him a Nobel Prize in 1912. Carrel made the artificial blood vessels using the glass and aluminium materials (Fung, 2013).

The current success in artificial blood vessels synthesis back dates to 1940s as a result of great work by Alexis. The surgical technique used by then before this great discovery was transplanting the arteries or veins from healthy donor to replace the damaged and sometimes the diseased vessels. This technique was faced with a lot of challenges, tissue rejection and development of arteriosclerosis being among them (Grassl, Barocas, & Bischof, 2004).

Materials such as vinyon, plastic Teflon and synthetic fibre Dacron were used to make artificial blood vessels and the results were encouraging. Vinyon was at some point tried on dogs and later human being in 1953 by Voorhees. The reports obtained from these studies suggested that synthetic blood vessels from these materials were rarely rejected by the human immune system. To add on the above advantage, these materials were readily available, cheap and durable. However, Dacron was found to have tendency of clogging by blood clot and research is still underway to design the interior walls of small synthetic Dacron blood vessels to eliminate the clot formations (Humphrey & Baroutaji, 2016).

The ideal artificial blood vessel graft should have the following characteristics

   I.   Compatible with the natural blood vessels

  II.   Lack the thrombogenicity associated with natural blood vessels

 III.   Should be resistance to infections

 IV.   Should have the ability to heal, contract and remodel.

  V.   Should be able to secrete normal blood vessels products

The above characters are necessary while structuring the artificial blood vessel scaffold from collagen or any biodegradable polymer. It should be noted that mechanical properties of artificial blood vessels is greatly enhanced by substances such as bioreactors that mimics the in vivo milieu of the blood vessels' cells by producing pulsatile flow (Fink, Sellborn, & Krettek, 2012).

2.2 Why do we Need them?

It has been shown by various surgical reports that transplanting ones' blood vessels to replace the diseased or damaged blood vessels lead to rejection. Moreover, transplantation required more than one surgeries, one for the healthy vessel to replace the unhealthy one and second surgery was removal of the unhealthy blood vessel (Song, Wang, Huang, & Tsung, 2014). This was consuming a lot of time, labor, and resources and also caused more harm to the patient as it led to more wounds that can easily be infected when good nursing practices are note observed. Surprising to note, these patients with poor circulation problems lacked suitable vessels that could be harvested for transplant. Therefore, discovery of artificial blood vessels synthesis and the material suitable for this was necessary and a big blessing (Grassl, Barocas, & Bischof, 2004).

The synthetic fabrics made of materials such as polyethylene and siliconized rubber had much advantage since they were porous and also flexible. Their porosity nature allowed permeation of some host blood cells into the material and this led to cell multiplication for natural tissue to emerge and the synthetic vessels get eliminated. This material seems have saved a lot of lives by preventing development of ischemia to heart parts supplied by affected coronary arteries which could have led to myocardial infarctions. The world has suffered a lot especially from complications related to rupture of coronary arteries due to high blood pressure, stenosis and aneurisms (Fink & Helen, 2009).

2.3 Additive Manufacturing

Rapid prototyping has been applied in industrial manufacturing of commercial products. It is a quick means to ensure mass production of substances which may be needed in bulk. It gives a prototype of the object required which finally will be modified to give the final product that looks almost similar to the natural one (Froelich et al., 2010). The whole process is known as 3D printing of additive manufacturing. The process involves fabricating the parts of the prototype product with additional materials hence additive approach. The principle behind this great technology is production of initial pro-product termed as prototype with aid of three dimensional computer aided design system. Once the pro-product is obtained, it is then remodeled without process planning by adding other substance to get a complete final product (Fink, Sellborn, & Krettek, 2012).

The other form of manufacturing artificial substances for commercial purposes in the industry is Rapid manufacturing. The method is mainly suitable for rapid processing of low volume products which are not required in bulk. The products of rapid manufacturing are commonly customized and have complex shapes. They are good for emergency. Rapid manufacturing have the following advantages: safer working environment, no need of tooling, assurance of dimensional accuracy and geometrical freedom. It should be noted that a lot of material are wasted in this technique making it more costly; it is slow and has a low surface finish (Fung, 2013).

2.4 3D Printing

Previous section had discussed the three dimensional printing technique and how it works. This section will reinforce the available mechanisms used to achieve production of plastic materials and other substances relevant to the topic of this paper. Three dimensional printing popularly known as additive manufacturing is a slow process that involves building product in layers. The initial step is always computer aided design sketch which is later developed to STL file which can be easily interpreted by the 3D printers (Froelich et al., 2010).

The available technologies using 3D printing system include stereo lithography (SLA), selective laser sintering (SLS), inject modeling (IJM), electronic beam melting (EBM), fused deposition modeling (FDM) among others. 3D printers can be dismantled to obtain five main parts, cartridge, coolant, electronics, print heads and finally built table. The 3D printers systems have been documented to posses the ability to build multi-material parts. Such examples include, Z-Corp 3D Printing and Object Inkjet Printing. Synthesis of artificial blood vessels requires high technology to ensure that the structure of the unhealthy natural blood vessels is replaced with almost similar artificial blood vessels to avoid the side effects that may result from too much deviation from the natural ones. For the purpose of this paper, only fused deposition modeling will be discussed in details (Humphrey & Baroutaji, 2016).

2.5 Fused Deposition Modeling

Fused deposition modeling as a technique of 3Dprinting technology can give the final products without synthesizing the intermediate termed prototype. The blood vessels are manufactured by extruding individual layers of the material using the nozzles and build them in layers as per the structures desired. It is capable of producing thermoplastic blood vessels with good mechanical strength that can withstand an elevated blood pressure (Dabiri, Schmid, & Tryggvason, 2014).

It should be noted that while using fused deposition modeling technique of 3D printing, the following parameters must be taken into consideration: the part build orientations, raster angles, thickness of the layers, road width and air gap. The extruded material is which is in thermoplastic state is drawn into head using a wheel drive attached in a motor moving horizontally hence molding the exact cross section of each layer (Fink, Sellborn, & Krettek, 2012). The plastic substance then hardens within a short period of time to form a union with previously molded layer. The substance holding the product is removed

6

by soaking in a solution. The build styles that can be applied in fused deposition modeling technique include solid normal, sparse-doubled and spars. The layers should be filled completely using solid materials to avoid vacuums. The raster width and raster angle should be as low as possible to enable more bonding which will enhance the strength of the final product (Dabiri, Schmid, & Tryggvason, 2014).

Fused deposition modeling is composed of three main stages which include creation of a prototype using a computer aided design file, heating the thermoplastic prototype to make it a semisolid and lastly printing the substance and socking to get rid of the supporting material (Snowhill & Silver, 2005). FDM technique is described to have anisotropic structures which other molded structures do not have. Because the technique is not hygroscopic, the final products are stable hence fit for use as artificial blood vessels to replace the diseased or damaged natural blood vessels. Fused deposition modeling makes use of materials such as polycarbonates, ABS, polyphenylsulfone, nylon and ULTEM9085. These materials yield the best structures that qualify to be used as artificial blood vessels (Dabiri, Schmid, & Tryggvason, 2014).

The benefits of artificial blood vessels manufactured via fused deposition modeling technique includes lack of toxicity, durability, cleanliness and environmental friendly, flexibility in design and high accuracy. However, the products and the technique face the following challenges: support of the product during synthesis is required, extrusion of the head must be kept moving without which the material will bump up, the process and the machine is costly (Dabiri, Schmid, & Tryggvason, 2014).

CHAPTER THREE

3.1 Previous Researches

The first in vitro arteries were constructed by Weinberge and Bellwere. The components of the artery were collagens and bovine aortic SMCs casted together to obtain an annular mold. The inner parts of the artery graft were made of bovine ECs while the outer surface fabricated with bovine adventitial fibroblasts. To maintain the mechanical strength of the artery, Weignberge and Bellwere used the Dacron mesh to furnish the walls of the artificial artery (Froelich et al., 2010). The researchers discovered that without the Dacron mesh, the synthesized artery would bursts at intraluminal pressure which is greater than 10mmHg. The more the Dacron mesh layers, the more strong the vessels and alternating the

Dacron mesh with collagen lattice was enough to make the artery much stronger and it would need a pressure of 120 to 180mmHg to burst (Grassl, Barocas, & Bischof, 2004).

Elastin absence in the circular smooth muscles lead to low density of the blood vessels and this was explained to offer poor support hence the vessels could easily rupture with minimal rise in pressure. The endothelium of such artificial blood vessel was capable of producing Von Willebrand's factor and prostacylin components; this made the scholars to venture in the study to find more about this kind of artificial blood vessel (Hoch, Tovar, & Borchers, 2014).

Hybrid medical tissues are prepared through molding of tubular glass using a cold solution of canine jugular SMC and type I collagen. The inner surface of the artificial blood vessel is molded using jugular EC and then the resulting graft implanted in canine vena cava. These kinds of artificial blood vessels rupture within few days because of lack of mechanical strength. The grafts can be meshed with Dacron to prevent tearing. This also boosts the shelf life of the product and also the stability within the place of transplant (Froelich et al., 2010). The complete remodeling of the artificial blood cells can be documented using a monolayer of endothelial cells laid longitudinally on the inner wall. The documentation of the smooth muscles circumferentially oriented at the subendothelial layer will also give some clue of the remodeling process. The remodeled structure of blood vessels are characterized by presence of fibroblast throughout the walls, presence of collagen fibril meshes at the intercellular space and sheet like lamellae in the endothelial layer (Fink & Helen, 2009).

To enhance mechanical strength of the artificial blood vessels, the smooth muscles orientations were to be changed in order to avoid stress and strains. The longitudinal forces are minimized by making the media to adhere to glass mandrel which is then broken up by regularly tapping the resulting blood vessel tube on a hard surface. The adventitia are also necessary in offering the strength, to boost their synthesis, a collagen gel with human dermal fibroblasts are added to stimulate the synthesis. At this step, the structure still cannot withstand the intraluminal pressure at normal physiological state hence need to strengthen further the artificial blood vessel. Umbilical vein endothelial cells are sometimes injected in the structure inner lining to boost the development of endothelial cells (Humphrey & Baroutaji, 2016).

Longitudinal orientation of the endothelial cells was found to enhance the strength of the artificial blood vessel hence offering the required stability to overcome the shear stress.

The combination of the longitudinal orientation of the endothelial cells and the circumferential orientation of the smooth muscle cells is works effectively to improve the vasomotor activities of the artificial blood vessels (Liao, 2011). Pulsatile flow on collagen based artificial blood vessels makes the endothelial cells to align parallel to the direction of blood flow as the smooth muscle cells aligns to parallel to the direction of stretch hence achieving the necessary mechanical strength to accommodate normal blood pressure. The nondegradable materials such as polyurethane grafts have been used together with collagen gel to embed the canine smooth muscle cells for achieving the strength. The intima is synthesized by seeding the autologous endothelial cells in the artificial blood vessels material media (Hoch, Tovar, & Borchers, 2014).

The application of pulsatile flow on an immature blood vessel manufactured by 3D Printing technology can lead to damages hence need to use a polyurethane scaffold. Alternatively, the scaffolding can be evaded by use of elastomeric silicon tube bioreactor. This will give a hybrid tissue made up of mainly type I collagen and bovine aortic smooth muscles which is eventually subjected to periodic cyclic inflation. These yield circumferentially aligned smooth muscles cells and collagen fibers (Sabiston, Sellke, Del, & Swanson, 2010).

A strong magnetic field has also been utilized in making the smooth muscles cells of the artificial blood vessels to take the orientation form desired.   The strong magnetic field is applied at the time when fibrous are synthesized leading to circumferential orientation of the collagen fibrils in the collagen scaffold. The smooth muscle cells are also trapped within the alignment hence taking the appropriate orientation. This technique is efficient as compared to other techniques used to orient the collagens and the smooth muscles cells. This is because the process is simple, reproducible and much stiffer and viscous media is formed which confers the mechanical strength (Grassl, Barocas, & Bischof, 2004).

Sleeve hybrid graft made of collagen layers of neonatal human dermal fibroblasts and inner support sleeve which is fabricated with type I collagen gel has been successful in ensuring the stability of the artificial blood vessels. The framework of this structure has been cross linked with glutaraldehyde in the presence of dehydrothermal and ultraviolet radiations. The resulting artificial blood vessel is suitable for transplant in animals but not human. This is because it has not yet been tested in human beings. The sleeves degrade as the cellular layer of the structure strengthens (Song, Wang, Huang, & Tsung, 2014)

Among the several biodegradable polymers that have been used to manufacture the artificial blood vessels, polyglycolic acid is most preferred. This is because it is porous; hence permitting the entry of nutrients once implanted hence allows neovascularizations to replace the artificial blood vessels with the natural blood vessels. The drawback of polyglycolic acid as a synthetic material for blood vessels is that the meshes rapidly get absorbed hence making the structure unable to bear the blood pressure. Good enough, the mechanical strength can be improved by modifying the physical stability of the polyglycolic acid at molecular level (Liao, 2011).

3.2 Artificial Vessels without the Use of the Scaffold

Human culture cells have the ability to undergo angiogenesis without the use of scaffolding techniques. Examples of such cells include the human umbilical cord veins smooth muscle cells and the skin fibroblasts. Culturing of these cells for 30 days would form cohesive cellular sheets with cells and extra cellular matrix that easily peels off from the culture flasks or support (Fung, 2013).

The resulting smooth muscles cells sheets are wrapped around the tubular support so as to obtain the media of the artificial blood vessel. The media matures after one week upon which the fibroblast sheets are rolled around the media to obtain the adventitia that will signal the synthesis of intima. Maturation period of seven days is allowed after which the mandrel is removed followed by seeding of the lumen with human umbilical vein endothelial cells in order to grow for another one week (Patel, 2012).

Finally, a three layered artificial blood vessel that almost resembles the natural human artery is obtained. These blood vessels can withstand up to between 2093 to 3095mmHg hence more stable and durable. These blood vessels rarely neither degrade nor tear (Liao, 2011).

3.3 Current Status and Future Perspectives

Cases of coronary artery stenos and aneurisms have been reported to be on increase. This makes the constructions and synthesis of the artificial blood vessels and surgical replacement also to be at the lime light. The number of successful surgeries related to blood vessels replacement has been reported to increase because of the frequent development being carried out by various researchers to enhance the procedures and techniques applied. The reports indicate that more than 40% of children who have undergone reoperative

10

reconstructive cardiovascular surgery have come out successfully without any complications (Fung, 2013).

As much as the artificial blood vessels synthesis technology has advanced, still the mechanical properties similar to the natural blood vessels are still an illusion. However the artificial blood vessels have been reported that they do not necessarily need to have same mechanical properties as the natural blood vessels in order to function properly. The materials used are capable of remodeling, repair and allows the growth of natural blood vessels to replace the artificial blood vessels completely. The materials can also adapt the milieu of the vessel being replaced depending on whether it is a vein or an artery hence suitable for the purpose (Fink, Sellborn, & Krettek, 2012).

This form of advancement in hemodynamic medicine is faced with three main challenges outlined below:

1. The technology is not suitable in emergency cases because of time it takes to prepare the artificial blood vessel as the graft currently being used requires more time for investigation according the United State of America Food and Drug Authority (FDA) (Sabiston, Sellke, Del, & Swanson, 2010).

2. Most of the biodegradable materials currently being used to manufacture the artificial blood vessels had already received approval from the FDA however, biopolymer are considered the best to make the vascular conduits hence this is a step backward as more time will be needed to convince the FDA to change the standards (Patel, 2012).

3. The chances of infections are much higher as a result of prolonged duration of culture. Also, the prolonged duration makes the cost of manpower, materials and the equipments go up (Sebastian E. Dunda et al., 2012).

Alternative cells such as endothelial progenitor cells (EPCs) are used to in cardiovascular tissue engineering due to the fact that they can differentiate into endothelial cells and also that they are found at the peripheral blood hence well distributed in the body. The tissue plasminogen activator production by endothelial progenitor cells match that of endothelial cells hence EPCs have the capability of preventing clot formation during blood vessels replacement. The resulting blood vessels from EPCs have the ability to undergo rapid regeneration, contraction and intrinsic nitric oxide induced vasodilatation (Hoch, Tovar, & Borchers, 2014).

11

Smooth muscles progenitor cells (SPCs) have been found to be effective in constructing artificial blood vessels. These cells have more integrin alpha 5beta 1 expression. This is essential to allow fibronectin attachment and other adhesive matrices (Patel, 2012). The combination of tissue engineering and gene therapy in developing artificial blood vessels will lead to synthesis of a more effective structure that will almost compare with the natural in terms of physiological and anatomical features. For example of combining the smooth muscle cells and tropoelastin gene will give a structure with more elastin hence improved contractility and relaxation. A combination of cyclic GMP dependent protein kinase gene with a smooth muscle cells give rise to a blood vessel with enhanced contractile characteristics (Froelich et al., 2010).

CHAPTER FOUR

4.1 Coronary Heart Disease

Heart muscles commonly known as cardiac muscles are supplied by coronary blood vessels which branches directly from the oater. These blood vessels supply the nutrients and oxygen to the cardiac muscles for normal physiological functions which is pumping the blood. Coronary arteries blockade is not a new term as most case have been registered. The plaques in the blood circulating in the blood have been commonly found to be lodged in these blood vessels making the vessels to clock (Cohn, 2012). The plaque can be part of a clot, degraded tissues and low density lipoproteins. The resulting condition is known as atherosclerosis. Coronary heart disease (CHD) is characterized by development of this plaque in the vessels supplying the heart muscles with blood leading to inadequate or total lack of nutrients and oxygen to cardiac muscles. This can lead to hypoxia or myocardial infarction. The development of hypoxia will lead to subsequent heart attack and failure to restore the blood supply to the cells will lead to death of section of the heart which is a serious health problem that has culminated to deaths (Grassl, Barocas, & Bischof, 2004).

Heart failure occurs when the heart can no longer pump enough blood to the body to meet the respiratory needs and other metabolic needs. It should not be forgotten than arrhythmias can develop in this case due to irregular generation of action potential or impulses by the heart pacemakers such as sinoartrial node (SAN), artrioventricular node (AVN) and purkinje fibers (Rezai, Podor, & McManus, 2004).

Ischemic heart disease is common in Australia and the most common forms are heart attacks and angina. Heart attack as described above occurs as a result of complete cut of blood supply to the heart muscles hence impairing the normal functioning of the myocardial cells (Sabiston, Sellke, Del, & Swanson, 2010). On the hand, angina is a chronic episodic heart condition characterized by sharp pain on the left chest and back running to the left index finger. It occurs due to periodic temporary block of the coronary arteries supplying the heart muscle cells. Angina is not life threatening as compared to heart attack (Fink & Helen, 2009).

In Australia, cardiovascular diseases have been recorded to be the most cause of death with the case being 43,946 deaths per year as per the statistics of 2012. The major risk factors associated to cardiovascular diseases in Australia include smoking of tobacco, excessive alcohol consumption, sedentary living, and poor dieting. These factors are termed modifiable factors since they can be controlled by an individual (Fung, 2013). Some risks factors are beyond an individual control and are termed non modifiable risks factors. Such factors include the age, genetic makeup, gender, and ethnicity. High blood pressure, high cholesterol levels in circulation and overweight or obesity are also risk factors classified as biomedical factors. Other chronic diseases such as kidney failures and diabetes have been found to have links with development of cardiovascular diseases however the mechanisms cannot be explained well (Sebastian E. Dunda et al., 2012).

4.2 Treatment or Management of Cardiovascular Disease

The treatments options available for cardiovascular diseases vary and depends upon the cause or the implicated risks factors. Before initiating the treatment or management plan, first determine the cause and other conditions that exacerbate the condition identified. The treatment should be directed towards the identified risk factor and the acerbating conditions (Fink & Helen, 2009). The modifiable risks factors are the most easy to deal with as compared to non-modifiable factors and biomedical factors. The treatment options ranges from surgical, use of drugs and dieting. The best option is always preferred. The purpose of this chapter is to address the surgical methods of tackling the coronary heart disease that involved damage or diseased coronary artery. The chapter will look at the artificial blood vessels, their construction, tissue engineering of the blood vessels, biomaterial scaffolding and finalize with mechanical properties (Froelich et al., 2010).

4.3 Artificial Blood Vessels

Artificial blood vessels are preferred in replacement of diseased or damaged coronary arteries because of their compatibility with the natural blood vessels, resistance to infection, ability to heal faster, lack of thrombogenicity, the ability to remodel, and ability to secrete normal blood vessels products (Wilson, 2011). The artificial blood vessels are made up of three fundamentals: the scaffolding structure which is either collagen made or biodegradable polymer, vascular cells and a milieu that will favor the synthesis and development of the artificial blood vessels. The strength and stability of the blood vessel artificially made will depend with the technique used and the material selected to make the blood vessels (Fink, Sellborn, & Krettek, 2012). The bioreactors that mimic the in vivo milieu of the vascular cells are employed to generate pulsatile flow which will orient the smooth muscles cells and the collagens appropriately to get the stable and strong orientation of the artificial blood vessels (Snowhill & Silver, 2005).

The tissue engineers have made progress in tackling the issue of artificial blood vessels instability and weaknesses. They have used materials that have the ability to adjust to the hemodynamic changes and chemical stimuli that may trigger the immunological reactions. Wounds with large arterial diameters have been effectively managed using these synthetic grafts. However, the uses in arteries with smaller diameter have not been good (Fink & Helen, 2009).

4.4 Construction of an Artificial Vessel

Construction of the artificial blood vessels requires structural scaffold, cells and conducive environment. These are the basic elements that should be guaranteed for quality artificial blood vessels. Scaffold structure offers the temporary framework to support the media from which the blood vessels tissues will grow and also offers the tube like shape of blood vessels until the cells being cultured produce their own extracellular matrix (Fink & Helen, 2009). The scaffold should be made of collagen matrices and biodegradable polymers. Tubular mandrels have been utilized as an alternative technique to support the artificial blood vessels being synthesized in vitro as a modified scaffold structure. The mandrel tube is usually removed at long last before the resulting blood vessel is implanted to the damaged or diseased blood vessel. The scaffold is seeded with smooth muscles cells, fibroblasts and endothelial cells to improve the mechanical properties of the artificial blood vessels (Fink, Sellborn, & Krettek, 2012).

4.5 Tissue Engineering of Blood Vessels

Tissue engineering aims to boost the biological properties of the artificial blood vessels to make it look similar to biological ones. Engineering of the blood vessels tissues involves application of 3D scaffolding for maintenance of physical strengths. The synthetic polymers and natural materials like collagen and elastin have been employed in tissue engineering of blood vessels to mimic the extracellular cell matrix of the cells and organize them until they mature (Froelich et al., 2010). Bioinert materials such as Teflon and silicon were found to illicit least immune response, and still allowed permeation of the blood tissues hence allowing re-growth of the replaced blood vessels as the artificial vessels degrade. However, the Teflon and the silicone materials were not initially meant for medical purpose as they have some underlying disadvantages that make them unfit. This has made the blood tissues engineers to continue with further researches so as to come up with the best material to be used in making the artificial blood vessels. The aim is to obtain materials that will allow synthesis of blood vessels which can allow cell to cell communications and able to carry out functions such control of tissue formation and cell growth (Grassl, Barocas, & Bischof, 2004).

Tissue engineering of the blood vessels may cause measurable contractions to the pharmacological agents such as serotonin, endothelial-1 and prostaglandin F_{2a} which are contained in well differentiated smooth muscles cells. These pharmacological agents are well expressed in caponins and myosins heavy chains. The experiment that was done on the swine pigs gave positive results as the transplanted saphenous artery did not developed any thrombi. This shows that tissue engineering is gaining grounds in artificial blood vessels synthesis to come up with the best alternative (Patel, 2012). The thrombogenicity of decellurized xenogenic vessels is mainly caused by lack of endothelial cells lining the interior wall the artificial blood vessels. The tissue engineers have worked hard to eliminate the problem through reseeding decedualized porcine aortas with initially expanded human endothelial cells and the harvested myofibroblasts. Finally, ex vivo decidualization of the vascular matrix scaffold have been discussed as the best alternative solution to thrombogenicity , immunological rejection and prolonged healing time. The long healing time is the attributed to cryopreserve or glutaraldehyde fixed grafts (Rezai, Podor, & McManus, 2004).

4 6 Biomaterial Scaffolds

Scaffold structures require specific materials which will allow efficient modeling of the artificial blood vessels in the support structure. The materials used should be able to augment, replace or perform the functions of the natural blood vessels. These materials can either be manmade or natural and can be either be the whole or just a section of the living structure. The properties of the biomaterial to be used depend on the cell type, implantation site and the strategy for tissue formation (Hoch, Tovar, & Borchers, 2014). The requirement for the biomaterial used for scaffolding is that it should be free from microorganisms that can lead to infections, it should be compatible with the natural blood vessels and other materials used in synthesis and finally it should not cause hypersensitivity reactions to the receiver. The material physical strength should be enough to handle the physiological blood pressure of the individual upon replacement surgery (Froelich et al., 2010).

Construction of the vascular grafts through seeding process has been facilitated using various biomaterials. The extracellular cell matrices and electrospun synthetic polymers have been used to successfully achieve the seeding process. Electrospinning technique is used to create nano fibre which will adhere to the scaffold framework well (Grassl, Barocas, & Bischof, 2004).

4.7 Mechanical Properties

Mechanical and physical properties of the artificial blood vessels are essential in determining the site of surgical replacement. A larger artery which carries blood at an elevated pressure requires that much stronger artificial blood vessels in order to withstand the pressure. The parameters that determine the mechanical strength and physical properties are burst strength, tensile strength and compliance (Wilson, 2011). BC films were found to be extremely stronger because of their crystalline nature and the planer kind of orientation of the ribbons in addition to their complex networking. The mechanical properties of BC film is comparable to mechanical properties of the big carotid arteries and its compliance looks that of native artery making it of more advantages as compared to materials. The culture media can be modified to enhance the density of the resultant arterial blood vessel which will improve the mechanical strength (Froelich et al., 2010).

References

Cohn, L. H. (2012). *Cardiac surgery in the adult.* New York: McGraw-Hill Medical.

Dabiri, S., Schmid, S., & Tryggvason, G. (2014). Fully Resolved Numerical Simulations of Fused Deposition Modeling. *Volume 2: Processing.* doi:10.1115/msec2014-4107

Fink, & Helen. (2009). *Artificial blood vessels - Studies on endothelial cell and blood interactions with bacterial cellulose.*

Fink, H., Sellborn, A., & Krettek, A. (2012). Molecular Understanding of Endothelial Cell and Blood Interactions with Bacterial Cellulose: Novel Opportunities for Artificial Blood Vessels. *Atherogenesis.* doi:10.5772/25365

Froelich, K., Pueschel, R., Birner, M., Kindermann, J., Hackenberg, S., Kleinsasser, N., ... Staudenmaier, R. (2010). Optimization of Fibrinogen Isolation for Manufacturing Autologous Fibrin Glue for Use as Scaffold in Tissue Engineering. *Artificial Cells, Blood Substitutes, and Biotechnology, 38*(3), 143-149. doi:10.3109/10731191003680748

Fung, Y. (2013). Mechanical Properties and Active Remodeling of Blood Vessels. *Biomechanics,* 321-391. doi:10.1007/978-1-4757-2257-4_8

Grassl, E. D., Barocas, V. H., & Bischof, J. C. (2004). Effects of Freezing on the Mechanical Properties of Blood Vessels. *Heat Transfer, Volume 1.* doi:10.1115/imece2004-60244

Hoch, E., Tovar, G. E., & Borchers, K. (2014). Bioprinting of artificial blood vessels: current approaches towards a demanding goal. *European Journal of Cardio-Thoracic Surgery, 46*(5), 767-778. doi:10.1093/ejcts/ezu242

Humphrey, J., & Baroutaji, A. (2016). Blood Vessels, Mechanical and Physical Properties of. *Reference Module in Materials Science and Materials Engineering.* doi:10.1016/b978-0-12-803581-8.02254-2

Liao, K. (2011). Surgical Treatment of Coronary Artery Disease. *Coronary Heart Disease,* 405-422. doi:10.1007/978-1-4614-1475-9_22

Patel, C. (2012). Primary Prevention of Coronary Heart Disease. *Behavioral Treatment of Disease,* 23-41. doi:10.1007/978-1-4613-3548-1_3

Rezai, N., Podor, T. J., & McManus, B. M. (2004). Bone Marrow Cells in the Repair and Modulation of Heart and Blood Vessels: Emerging Opportunities in Native and Engineered Tissue and Biomechanical Materials. *Artificial Organs, 28*(2), 142-151. doi:10.1111/j.1525-1594.2004.47334.x

Sabiston, D. C., Sellke, F. W., Del, N. P., & Swanson, S. J. (2010). *Sabiston & Spencer surgery of the chest*. Philadelphia: Saunders/Elsevier.

Sebastian E. Dunda, T. Schriever, C. Rosen, C. Opländer, R. H. Tolba, S. Diamantouros, ... N. Pallua. (2012). *A New Approach of In Vivo Musculoskeletal Tissue Engineering Using the Epigastric Artery as Central Core Vessel of a 3-Dimensional Construct*. Plastic Surgery International.

Snowhill, P. B., & Silver, F. H. (2005). A Mechanical Model of Porcine Vascular Tissues-Part II: Stress–Strain and Mechanical Properties of Juvenile Porcine Blood Vessels. *Cardiovascular Engineering*, *5*(4), 157-169. doi:10.1007/s10558-005-9070-1

Song, S., Wang, A., Huang, Q., & Tsung, F. (2014). Shape deviation modeling for fused deposition modeling processes. *2014 IEEE International Conference on Automation Science and Engineering (CASE)*. doi:10.1109/coase.2014.6899411

Wilson, R. F. (2011). Transcatheter Treatment of Coronary Artery Disease. *Coronary Heart Disease*, 389-403. doi:10.1007/978-1-4614-1475-9_21